S0-BRP-625

Not a Butterfly Alphabet Book

It's about time moths had their own book!

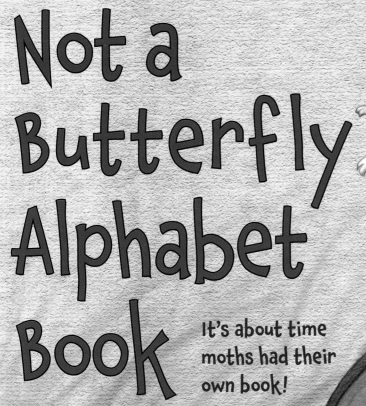

Jerry Pallotta
Shennen Bersani

Charlesbridge

To my Seaside (Oregon) to Scituate (Massachusetts) cross-country biking buddies:
Billy Nicholson, Dennis Sullivan, Frank Mastrocola, and the Blonde Woolly Bear
Caterpillar that rode with us for one hundred miles—J. P.

To Lou Marcoccio, with love—S. B.

Text copyright © 2019 by Jerry Pallotta
Illustrations copyright © 2019 by Shennen Bersani
All rights reserved, including the right of reproduction in whole or in part in any form.
Charlesbridge and colophon are registered trademarks of Charlesbridge Publishing, Inc.

At the time of publication, all URLs printed in this book were accurate and active.
Charlesbridge, the author, and the illustrator are not responsible for the content
or accessibility of any website.

Published by Charlesbridge
85 Main Street
Watertown, MA 02472
(617) 926-0329
www.charlesbridge.com

Printed in China
(hc) 10 9 8 7 6 5 4 3 2 1
(sc) 10 9 8 7 6 5 4 3 2 1

Library of Congress Cataloging-in-Publication Data
Names: Pallotta, Jerry, author. | Bersani, Shennen, illustrator.
Title: Not a butterfly alphabet book / Jerry Pallotta ; illustrated by Shennen Bersani.
Description: Watertown, MA : Charlesbridge, 2019.
Identifiers: LCCN 2018042980 (print) | LCCN 2018047421 (ebook) | ISBN 9781632898715
(ebook) | ISBN 9781632898722 (ebook pdf) | ISBN 9781580896894 (reinforced for library
use) | ISBN 9781580896900 (softcover)
Subjects: LCSH: Moths—Juvenile literature. | Alphabet books—Juvenile literature.
Classification: LCC QL544.2 (ebook) | LCC QL544.2 .P335 2019 (print) | DDC 595.78—dc23
LC record available at https://lccn.loc.gov/2018042980

Illustrations done in Prismacolor pencils on Arches watercolor paper
    and manipulated in Photoshop
Display type set in Chaloops, designed by Chank
Text type set in Digby, designed by Atlantic Fonts
Color separations by Colourscan Print Co Pte Ltd, Singapore
Printed by 1010 Printing International Limited in Huizhou, Guangdong, China
Production supervision by Brian G. Walker
Designed by Susan Mallory Sherman

A is for Atlas Moth

Don't even think about calling this creature a butterfly! This is a moth. The Atlas Moth is the largest moth in the world. Wingtip to wingtip, this moth can be as wide as this page.

B is for Bella Moth

The word *bella* means "beautiful." Most people think butterflies are more colorful than moths and have more intricate and interesting wings. It's not true! Moths are spectacular, too.

C is for Cow Moth

Butterflies and moths land differently. Most butterflies land with their wings folded up. Moths land with their wings spread out. Cow Moth, we want to say one thing. *Moo!*

D is for Diamond Moth

This moth looks like it belongs in a jet book.
It is the same shape as a Delta Dart fighter jet.
Moths are insects, and insects have six legs.
A jet is an aircraft and does not have legs.
It has landing gear instead.

E is for Emperor Moth

My, what lovely fake eyes you have! Since birds, tiny mammals, and small reptiles enjoy eating moths, these pretend eyes are perfect for scaring predators away. If a predator catches the moth, it usually eats the body and leaves the wings.

F is for Fly Moth

Nature is truly amazing! This moth has upper wings that mimic red-eyed flies eating bird poop. Some moths have wing designs that mimic eyes, sticks, and leaves. This moth is also called a Mural Moth.

G is for Green Lips Moth

While walking through the rain forest, look for the Green Lips Moth. No kissing allowed! The Green Lips Moth could be mistaken for a green leaf. It is also known as a Giant Silkworm Moth.

It looks like a hummingbird.
It flies like a hummingbird.
It sounds like a hummingbird.
It eats like a hummingbird.
Don't be fooled! It's a moth.

H could have also been for Hercules Moth.
Some people think that the Hercules
Moth is the largest moth on earth.
Others disagree and say that the
Atlas Moth is the largest. Oh no!
We have a controversy.

H is for Hummingbird Moth

I is for Imperial Moth

This yellow moth is camouflaged when sitting on a yellow flower. A scientist who collects and studies butterflies and moths is called a lepidopterist. *Lepidoptera* is a Latin word that means "scale wing."

What do butterflies, moths, reptiles, most fish, and two mammals have in common? Scales!

LIZARD

This orange gecko is a lizard that has scales.

PANGOLIN

A pangolin is a mammal with scales made of keratin. Keratin is what our nails and hair are made of.

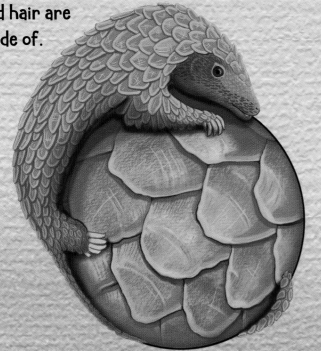

MOTH

This is a close-up of a moth wing. Butterflies and moths have delicate wings with tiny scales. If you touch them, the scales wipe off and look like dust on your fingers.

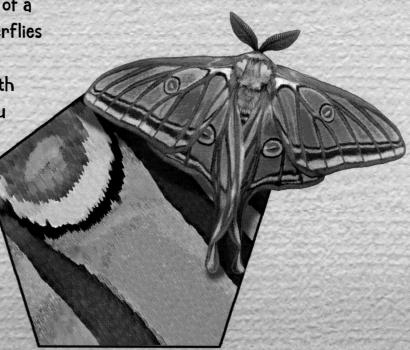

SNAKE
Snakes are reptiles that have scales on their skin.

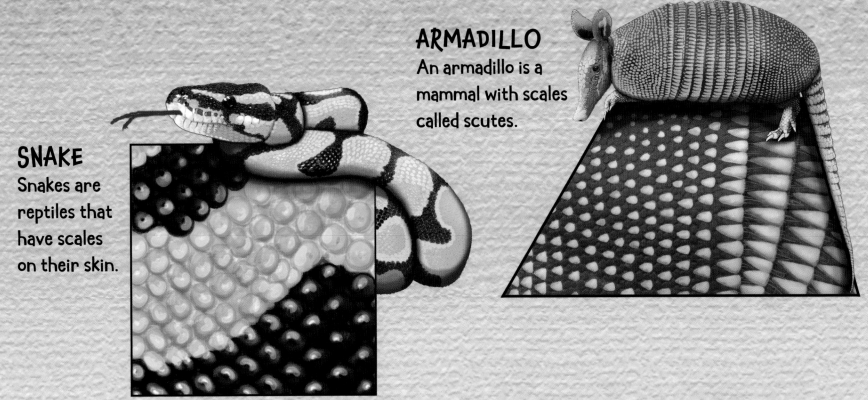

ARMADILLO
An armadillo is a mammal with scales called scutes.

FISH
This clown fish is scaly, too.

BUTTERFLIES
The scales on butterflies' and moths' wings overlap like shingles.

J is for Jersey Tiger Moth

hindwing

forewing

abdomen

thorax

head

eye

proboscis

antenna

leg

A Jersey Tiger Moth has hindwings that are an entirely different color than its forewings. This is a moth anatomy lesson. Moths have two antennae, two eyes, a head, a proboscis, a thorax, an abdomen, a pair of forewings, a pair of hindwings, and six legs.

K is for Kentish Glory Moth

Sorry, penguins! Moths live on every continent except Antarctica. The Kentish Glory Moth is found only in Scotland.

L is for Leopard Moth

Leopards aren't the only animals with spots. Another good name for this moth could have been Dalmatian Moth.

The long hindwings of the Moon Moth are similar to those of the Luna Moth on the title page of this book. When long-tailed moths fly, their tails create vibrations that distract their biggest predators—bats!

M is for Moon Moth

This moth is named after a spice. Do some research. Are there other moths named after spices or herbs? Is there a Parsley Moth? Sage Moth? Rosemary Moth? Thyme Moth?

N is for Nutmeg Moth

O is for Oakworm Moth

Cockroaches, sugar gliders, and skunks are creatures that shy away from light. Moths are the opposite. They are attracted to light. Moths will fly toward a lit candle, lantern, or porch light. Want to get rid of moths? Turn off the lights!

Butterflies usually fly during the day. Moths usually fly at night. The Pandora Sphinx Moth is most often seen flying at dusk.

P is for Pandora Sphinx Moth

Q is for Quaker Moth

Moths do not have teeth or fangs. They can't bite you, and they don't chew their food. Moths drink with their proboscis, which looks and works like a straw.

R is for Rosy Maple Moth

The Rosy Maple Moth looks like a tropical insect, but it is actually found in the East Coast forests of North America. It gets its name from eating mostly maple tree leaves. People enjoy maple flavor, too.

S is for Sheep Moth

Butterflies have straight antennae with a bump or a club at the tip. Moths have pointed antennae. Some moths, like these Sheep Moths, have feathered antennae.

The Tapestry Moth is also called the Carpet Moth. These pesky moths get their name because their larvae feed on carpets, fabric, wool, or even birds' nests. They can ruin clothing and furniture in your house. If this moth gives you trouble, call an exterminator!

T is for Tapestry Moth

U is for Underwing Moth

There is no such thing as a baby moth. When a moth comes out of its cocoon, it is already an adult. It is a baby during the caterpillar stage. Let's learn about the moth's life cycle on the next page.

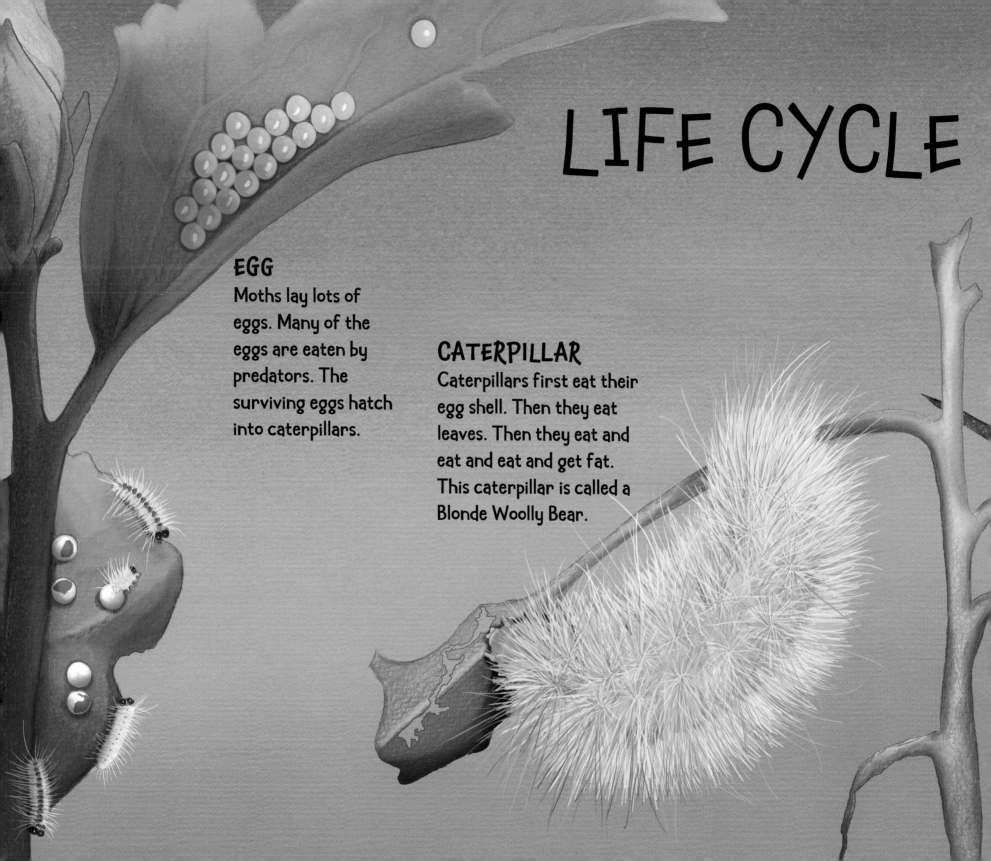

LIFE CYCLE

EGG
Moths lay lots of eggs. Many of the eggs are eaten by predators. The surviving eggs hatch into caterpillars.

CATERPILLAR
Caterpillars first eat their egg shell. Then they eat leaves. Then they eat and eat and eat and get fat. This caterpillar is called a Blonde Woolly Bear.

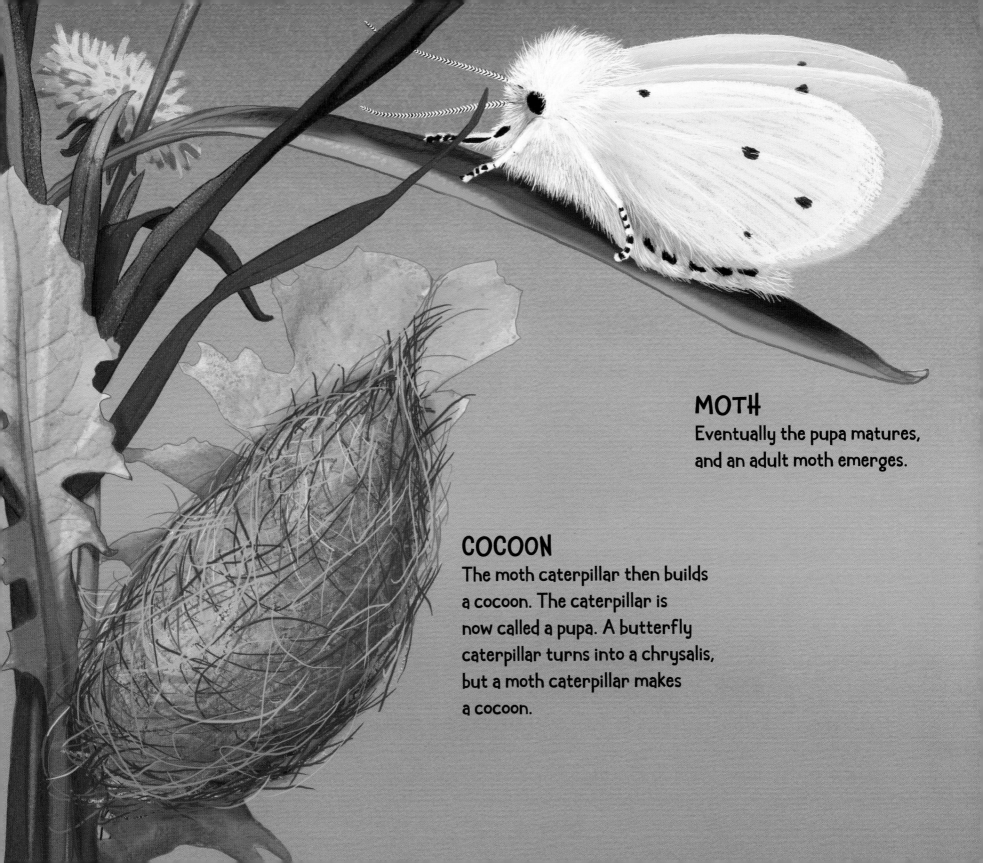

MOTH
Eventually the pupa matures, and an adult moth emerges.

COCOON
The moth caterpillar then builds a cocoon. The caterpillar is now called a pupa. A butterfly caterpillar turns into a chrysalis, but a moth caterpillar makes a cocoon.

V is for Venus Moth

This moth is fuzzy. It's a myth that moths are hairy and fuzzy and butterflies are not. Both moths and butterflies can be fuzzy wuzzy.

W is for Wasp Moth

Moths do not have claws or nails. A moth cannot scratch you, dig a hole, or get a manicure. How do moths defend themselves against predators? They fly erratically and are difficult to catch. This Wasp Moth's best defense is that it resembles a nasty stinging wasp. No one wants to mess with a wasp!

X is for Xestia Moth

Moths versus butterflies! There are ugly moths and ugly butterflies, and there are beautiful moths and beautiful butterflies. Scientists tell us there are 150,000 different species of moths. There are 20,000 species of butterflies. There are not many moths that begin with the letter X.

Y is for Yucca Moth

What a picky eater!
Picky! Picky! Picky!
The Yucca Moth likes to
visit only one type of plant—
the yucca. Moths prefer white
plants and flowers because they
are easier to see at night.

Z is for Zigzag Moth

Looking through this book, do you see a moth that might be a good design for a shirt, a coat, pants, or a scarf? Zigzag is a funky pattern to try out in your art class. Another word for zigzag is *chevron*.

Remember, this is NOT A BUTTERFLY ALPHABET BOOK! There are a zillion children's books about butterflies. This is a MOTH BOOK!